Creation Equation

By

Matt Schlidt

Dedication

I'd like to dedicate this book to my brothers, Tim and Joe Schlidt, and their belief in me as well as being people that I believe in too. My whole life they've been there for support, and they were the foundation of my interest in Science, Math, Business, Literature, and more. The love that comes from them is immense and I reciprocate my love to them to the fullest. That love is something I hope reaches every heart and engages every mind of every kind of reader and person, where we can hopefully create an optimal future through Elemental Frequencies that effectively utilize the Creation Equation.

About the Author

Matt Schlidt is a scientist, entrepreneur, and investor. Since his teenage years he's founded several companies and been a core employee for startups. Although Matt spent years in business, Science is his true passion.

Matt has been displaying this passion since Tonawanda Elementary, where he was taught the basics of Chemistry, Physics, Biology, and even Environmental Science. These served as the foundation of his Science education, which continued into Pilgrim Park Middle School, and eventually Marquette University High School (MUHS). MUHS was one of the Schools where Matt initially got exposure to Science Labs and from the first Lab to his most recent experiences, Matt has always been captivated by the Scientific potential behind experiments.

One experiment Matt received remarkable success with by building off of a phenomenal scientist named Rhonda Megal was with the study of Planaria and electromagnetic radiation (EMR), which proved EMR is cancer-causing. Matt received many awards for this experiment, such as the MU Alpha Theta Award, Badger State Science & Engineering Fair Award, an American Nuclear Society Award, and the Society for Optical Engineering Award. These Awards were given to further show the importance of the significant breakthrough, which supports the potential for new technology.

Experiments are crucial to education, but mostly important in discovery by pulling together test samples and tools to measure outcomes. Matt is a believer that Optimization can occur through Scientific Exploration and he has embraced this his entire life. Matt has been searching for the right answers in Science and Mathematics, but mostly experimentally.

Matt hopes that the breakthrough answers he has provided through Science and Math can serve society in the most beneficial way possible. What he discovered is what you believe you can achieve, making the possible probable.

Contents

Foundational Introduction

What if God created our Universe and you were able to assist in creating a new one? Well, God did, and you are to be a key to unlocking the door to a new Universe. As a matter of fact, you already were fundamental in creating a new Universe amongst us and that can serve all beings infinitely. You were behind the wormhole to our lives and existence, which was created by the energy of life that can be broken down into the following Creation Equation:

$$E = (MaTt)s^2$$

This is Energy (E), which is equivalent to matter (M) multiplied by area (a), time (T), and thoughts (t), as well as spirit (s), which are the frequencies of all atoms that travel at the speed of light based on the strength and belief in the spirit they possess. Energy comprises all. Matter makes up the mass of each atom. Area is the space all around us and on our planet, within the solar system, and the greater Universe. Time is what enables life and all matter to evolve with love, while thoughts and spirits define not only the present, but the future too.

The frequency exhibited by a spirit emits thoughts that shape all that we know. Light travels at the fastest proven speed today, but I believe that spirits and thoughts are equivalent and make up what can be considered the speed of light. Creation foundationally began with spirits and thoughts traveling at the speed of light to lay the foundation for light and the beginning of the Universe that would lead to matter and all of Earth as well as the area before and after it. The equation that can be fundamental to representing spirit, thought, and the speed of light is represented below:

$$s \times t = c$$

Elemental Frequency was foundational to the creation of a new Universe. The Frequency and Creation Equation that I developed can now be considered more of a Factor. Especially when the research is done to prove it, but this Theory is significantly founded on the unification of the Special and General Theories of Relativity, as well as Quantum Mechanics. I call it a factor because it was key to opening the door to a new Universe and enables us to open the door to more Universes, galaxies, solar systems, and planets in the future. The theories above were established by Nobel Prize-winning physicist Albert Einstein. In addition to these two foundational theories, Quantum Mechanics—pioneered by Max Planck—further unifies many of these principles alongside Elemental Frequency Theory, which may at times be expressed as Universal Frequency Theory.

The Special Theory of Relativity can be explained by the interplay of space and time. It states that two observers moving at different speeds will experience space and time differently.

Ultimately, this leads to time dilation, which means that time will pass more slowly for those moving at speeds close to the speed of light. Frequency Theory expands on this through the discovery that waves of energy, such as photons in motion, cause space and time to appear to move at different speeds due to Matter, Area, Time, Thought, and Spirit (Frequency)—also known as (MaTt)s all playing a foundational factor.

These waves can push creation beyond its current boundaries and onto a new frontier or collapse space, time, and matter to form a new Universe. This is only possible when the matter of a Universe has evolved to comprise life, and all life possesses a soul endowed with thoughts. Waves with even shorter wavelengths than gamma waves create universes, while gamma waves and even longer-wavelength waves form stars that lead to galaxies and solar systems, which then comprise those universes.

The General Theory of Relativity, another theory founded by Einstein, states that mass and energy lead to the prediction of gravitational effects. This further explains phenomena such as the bending of light, time dilation in strong gravitational fields, and the motion of objects in gravitational fields like stars, which can be created by humanity in case ours ceases to survive as long as we can on Earth and in our solar system.

Einstein's theory better explains how light creates Universes, galaxies, solar systems, and stars within them. This is because the first universal element, hydrogen—or, as many know it as a factor of light—condenses due to gravity. When God created light to peer through a wormhole the gravity condenses light in open space and the hydrogen pulls and pairs elements together to form light-dust that enable hydrogen and eventually helium.

Elements expressed their importance with Hydrogen and Helium creating bright stars, Oxygen creating the fuel for life, Nitrogen assisting in atmospheric development, and with Carbon being the backbone of living beings. Following the formation of a star, the light shines into the Universe and evolves elements to become resources on planets and, eventually—hopefully—life on a planet like Earth. Gravity and light are waves that are straight but curved in space-time-matter and can optimize frequency to better balance stars and planets.

Planck's theory, Quantum Mechanics, describes the behavior of matter and energy on the smallest scales, incorporating wave-particle duality, quantization, probabilistic outcomes, and the Heisenberg Uncertainty Principle. Wave-particle duality is foundational to quantum mechanics. It is fundamental in describing how subatomic particles exhibit both wave-like and particle-like characteristics. These wave-like and particle-like characteristics can achieve elemental alignment with their most ideal pair if the right thoughts and frequencies are exercised.

Quantization, another component of quantum mechanics, refers to the fact that the properties of particles and systems—such as matter, energy, and angular momentum, which describes the rotation of an object while considering its mass and rotational speed, can only take on discrete values.

Classical physics opposes and disagrees with quantization, as do I. Physics can largely justify that the discrete values of atoms can change due to subatomic exchanges. The vibrational frequency of each atom can be sensed on Earth and in the atmosphere. Following this scientific and spiritual sense, an atom can be driven closer to its optimal atomic match based on its atomic makeup and subatomic composition. This interaction of energy and vibrational frequency can atomically change molecules and elements to create better matter universally—to the point of perfection.

Copilot provides an interesting insight into quantum mechanics, stating: "Probabilistic outcomes refer to the idea that the exact state or behavior of a particle can't be determined with absolute certainty. Instead, we can only calculate the probability of finding a particle in a particular state or location. This is encapsulated in the concept of the wave function, which provides a probability distribution for a particle's position, momentum, and other properties."

While intriguing, I do not believe this insight on quantum mechanics is entirely true. Ultimately, we determine what happens to particles in the end—beginning with light and progressing to thought, which is driven by the matter that develops and evolves us as humans. Loving evolution ensures that we are guided toward the most positive and optimal outcomes. Our mind, body, and spirit can shape this process by utilizing the collective consciousness to unify Earth, the Universe, and every element within it.

Expanding on this, the Heisenberg Uncertainty Principle is a fundamental concept in quantum mechanics that states it is impossible to know both the exact position and exact momentum of a particle simultaneously. Elemental Frequency Theory proposes that all particles vibrate at specific frequencies, attracting or repelling one another due to their atomic composition and interactions with universal waves. These influences can be sensed or measured, expanding our understanding of the Universe by incorporating space, time, matter, and even thought—what some consider frequency. There exists an empty space that surrounds the Universe, which I call *NoMaTt*—meaning *No Matter, Area, Time, or Thought* exist in that area or spirit as well.

Elemental Frequency Theory can be summarized as follows: The frequencies at which we and the Universe's elements vibrate shape our past, present, and future. These frequencies are what I consider thoughts and every element thinks to a point. This process determines the well-being of our Solar System, Galaxy, and the greater Universe. By aligning every element, cell, and being with its ideal pair through biological, mental, physical, and spiritual energy, we can optimize the presence of *MaTt* within our Universe. Additionally, this may lead to the potential creation of a new Universe that resonates with a specific creative frequency. This new Universe could either flow back to the point of origin at the beginning of the Multiverse or emerge through a Big Crunch. The possibilities are vast—something only God can grant us.

Spirit and matter determine the present and future

of our Planet, Solar System, Galaxy, and Universe. Atomic composition yields the elements that construct our reality, enabling both physical structure and the expansion of our imagination. Time, Space, and Matter are the guiding forces that transition the present into a sustainable future. To optimize the development of our lives and the Universe, we must harness these fundamental components effectively on Earth.

Frequencies vibrate Time, Space, and Matter, utilizing relativity to guide elements. If Elemental Frequency Theory is practiced with one hundred percent effectiveness, it will allow life to thrive infinitely. Optimization through effective application can create even more Time, Space, and Matter in our Solar System and beyond. We can even use the frequencies of the poles to move planets—along with all their life and elements—by utilizing waves to push Solar Systems closer together at the speed of thought.

This is most likely unachievable until after a Nuclear Fusion Reactor has been built, along with the creation of other Hydrogen/Light Energy sources to develop LightNet. LightNet is a network built from light that will replace the Internet, radio, cable, cell networks, electricity, and more. Constructing LightNet is a massive undertaking, but it is certainly possible within the next 25 years. Achieving this by 2050 will enable Earth to attain sustainability for a long and prosperous future. Although significant challenges may arise in various areas, building the network from Hydrogen—which can be converted into light—may prove suitable for all of society.

A Fusion Reactor can be developed and placed in a safe and efficient location globally. Hydrogen energy can then be transmitted from the reactor's core as Gamma Waves, which will be converted into the Visible Light spectrum and distributed via transmission lines and cables to power Earth with clean energy. Proper cabling and transmission infrastructure will be necessary to ensure that light is efficiently delivered from point to point, effectively sustaining the global economy. Ultimately, this will create the most advanced economic system Earth has ever seen, enabling us not just to survive, but to thrive for the time we are granted by God and our nearest star, the Sun. Infinity is still a possibility; let's seize what lies before us.

We are at a point where acknowledging that electricity and the networks we've built are harmful to all life. Based on my study, *The Effects of Electromagnetic Fields on Life*, this fact is evident.

Electromagnetic Radiation doesn't just cause cancer, when it is emitted on Earth, the energy fires back at the Sun like a gun, injuring it with each shot from our current electric network. The Internet, Radio, Cable, Electricity, and even more networks formed by humans all create toxic waves, decreasing our time to survive on Earth as we hurt the environment and all living beings in the Solar System. This brings us to our greatest threat, the Sun exploding, which could occur eventually, but we are decreasing the time we have to survive with each emission of Electromagnetic Radiation as it attacks the Sun.

When the Sun explodes, creating a black hole, all Electromagnetic Radiation on Earth will shoot into the Universe at other Stars and Planets, threatening their existence too. This could cause them to explode as well, creating what could be hell into a new heaven that makes the most of what we have into the future of a galaxy that lays a framework to a new Universe focused on helping more than hurting all energy going into the future.

Moving forward, we must seize the opportunity to transition the economy to function based on light rather than electricity. In many ways, we already do, but establishing a Light Network powered by a Creation Reactor, along with Light & Hydrogen Power Plants utilizing freshwater from desalination, is crucial to generating substantial, attainable, and sustainable energy. This presents our greatest chance to combat global warming and climate change. Light and Hydrogen as our energy sources should be the core of our economy. If not, we must cautiously use technology so that we ensure we survive for as long as possible, while not hurting the one Sun we have.

By utilizing fundamental elements like hydrogen and oxygen within optimal magnetic frequencies, we can create an energy source that remains sustainable indefinitely—depending on how intelligently we act to build the future. I believe we can largely achieve this transition within the next decade, but certainly by 2050. A 25-year timeline is a reasonable target both nationally and internationally.

We can begin by completing the development of a Fusion Reactor, the ultimate combination of elements. Then, by aligning this system with the poles, we can ensure the global transmission of light while also optimizing the Earth's natural elements.

Magnetism plays a crucial role in supporting life on our planet, making it logical to establish its optimal usage. Since all atoms have magnetic charges, we must maximize their potential within the poles. This will empower Earth to better manage global warming and sustain life for as long as possible. Let's optimize the present so we can build a better future—ensuring that we do not face the early destruction of the planet we call home.

Frequency Factor

Atoms vibrate according to their composition and their relationship to one another. All atoms and elements matter. What matters most to life and the Universe is light, which is a form of love. Light makes up a component of every type of element, which is why they are visible to life. This is what supports the birth of every Universe we've ever had. The magnetism of elements and their path of existence through space and time cause them to evolve due to several forces. That's why, as a Planet and a Nation, it is important to engage with the Poles. By doing so, we can more efficiently exercise all matter on Earth as we evolve closer to a Universal Utopia.

Gravity is a factor that guides atoms closer together. When an element approaches another, it exposes all magnetic charges in its core, as well as the electrons that rotate around them, to the other. The electrons can cause the protons to transfer their primary positioning to a secondary positioning in another element. This interaction evolves the area as well as the existence of those elements, transitioning the Universe—hopefully for the better—while also potentially changing the atomic composition.

Eventually, after atoms form a star, they enable planets to form in galaxies and solar systems. These galaxies and solar systems will also form planets, some of which may have magnetic poles capable of supporting life. Life is crucial to planetary, chemical, and biological evolution. Evolution will drive the development of the solar system forward. The thoughts that life generates will exercise the light within neural networks, both spiritually and mentally, to maximize positive evolution—not just the evolution of life but also the evolution of elements and molecules.

Our Universe began with light. Light guided all matter forward. Eventually, this progress curved the Universe, as Einstein observed. The curvature of (MaTt)s allowed us to increase the amount of time ahead of us. It seems that by matching elements together correctly, we can achieve infinite time. However, if we mismatch or misuse elements and resources, we shorten the time, space, matter, thoughts, and spirit, we have Universally. It is our duty to practice what is best for all of us.

Every life has a unique perspective, but there is only one truth, regardless of individual belief. The importance of beliefs comes from the trust between you, me, and all of us. Typically, the most trusted are the most believed. Honesty in science, finance, and religion is extremely important to humanity because it either pushes us forward or holds us back.

Everyone wants to move forward in life, and many times, this happens through discovery. I believe the Elemental Frequency Theory (EFT) is one of the most important theories ever discovered. It builds upon past and current scientific truths and aligns with a holistic way of life. This way of life grants us better health, mentality, and wholeness within the soul. It also aligns us with all that exists. Our existence has the potential to create a Universal Utopia with Earth—depending on how we choose to live it and how we use it in the future.

Time has the potential to be infinite. That can be achieved, but it doesn't mean it's easy—especially considering that infinity requires an unlimited expansion of $(MaTt)s^2$. The right frequency can create matter, so it's important how you live and treat others physically, mentally, and spiritually. Evolution occurs constantly and thrives best with love, so use your heart in relationships, and you will find a solution to evolution. There's a reason you can find the word "love" in "evolve." Remember, don't forget to use your heart with others.

At the beginning of time for our Universe, light expanded space as it shone brightly ahead. Gravity pulled together matter to create stars and planets. The interaction of these stars and planets, along with other elements from the past to the present, has gifted us a present moment so abundant that we can work together to give the creator of the Universe an infinite future.

Matter is almost everything—except space and time. No matter how you look at it, we must make the most of the matter we have. We must also create more in the present and into the future. It has been said that matter cannot be created or destroyed, but if we vibrate the frequency of elements in just the right way, we can yield more matter in the future. The elements can come from the origin of all matter, which is the beginning point of our Universe and the place where MaTt expanded from.

From this origin point, we'll pull matter from another Universe and expand space-time within our Universe, which I call "MaTts" (Matter, Area, Time, Thought, and Spirit). This expansion can continue infinitely into the future, gifting us unlimited resources. These resources can sustain us forever. Is forever a goal of yours? Your soul will live forever regardless, but we can achieve it in physical form, too. I would love to create a life that physically exists infinitely. I'm confident we can achieve this.

Let's make things better for you and me by first aligning our matter and optimizing our spirits, matching what's best for one another. From the smallest element to the most significant life on Earth and throughout the Universe, everything has a match. Let's match matter to create (MaTt)s forever into the future by mating and believing in spirit.

Ever felt someone's "vibe" and thought, "Wow! I really like how I feel around this person." That vibration between you and another means that you're better together. We almost always have the potential to be better together with every individual. However, it's important to understand that each person is best around one person of the opposite sex, rather than the rest. Meet your potential globally, and we can put your potential into practice. With practice, we can achieve universal perfection.

Although I believe we have the potential to put a Big Crunch or Freeze in place, my hunch on how the future will evolve is recognizable. With the death of more matter and thought, we will shift into a black hole within our solar system moving closer to a system-based singularity for all that makes up the planet we call home. Hopefully, we can birth a star by merging together gas and dust clouds in our greater Solar System. However, thankfully, what we know will suit us infinitely into the future of our universe. From this universe to the next, let's be at our best. Let's do that by optimizing space-time-matter in a way that allows the best to happen through our thoughts. Begin by matching the vibrational frequency of each element, and we can live forever. That much is evident to me. In time, you will see that truth, regardless of whether you believe it.

The Law of Conservation of Mass comes from Antoine Lavoisier. Lavoisier stated in the law that the mass of any one element at the beginning of a reaction will equal the mass of that element at the end of the reaction. However, when an element truly reacts with its optimal vibrational pair based on their frequencies, there can most certainly be an exchange of electrons, protons, or neutrons that changes the element's mass and atomic composition.

Given this, we can create matter through our mind, body, and spirit. Space, time, and matter are guided by elements and the cells that make up life. The spirit we possess can be better recognized by science and society moving forward. Once we learn the foundation of using our spirit to better guide the Universe, we will be better suited for life on Earth.

Vibrations tend to not remain constant, regardless of whether you are alive or not, they typically decay over time. They continue unchanged even when a black hole with a Universal collapse to a MaTts point is created. Alternatively, and more sustainably, a wormhole can be formed to act as a window to a new Universe. The black hole compresses everything into a single point, so dense that all elements combine to form the most reactive molecule that can eventually expand within the Universe through atomic alignment, also known as elemental pairing. This molecule is so reactive and packed with maTt that it has the potential to create and expand the current Universe through the pairing of Matter, Area, Time, and Thoughts from You, while utilizing Energy, and the Spirits of all. This is all part of the use of the equation:

$$E = (MaTt)s^2$$

It is crucial to carefully consider these ideas because when falsehoods guide our thought processes, space, time, and matter can be lost, causing the Universe to darken in its area. You, yourself, have the power to influence what happens to everything else. Manifest the most you can through your thoughts, words, and actions, which dictate your beliefs.

Practicing the Theory

Putting theory into practice involves using your mind, body, and soul wholly. You must feel the vibrational frequency of all others. For humans, this can mean feeling the sexual nature of the opposite sex. If you're a man, find the vibrational frequency that most closely resembles you in a woman and engage with her mentally, physically, and spiritually. If you're a woman, seek the opposite in a man. All life can be matched, from humans to ants. Men can also feel the nature of other men, just as women can feel women, but these beings cannot be matched in order to enhance the Universe.

Following alignment with your potential mate, match elements, molecules, and cells. Spiritually guide them toward where they are best suited. Religions have been exercising Elemental Frequency Theory for centuries. They've labeled it with different terms, but it is here to stay forever. If we can fully tap into the frequency of the Universe—which stands for "One Song"—we can create the best vibration possible.

Vibrations affect the resources of every nation.

This can better suit the candidates who lead nations, the citizens who live within them, and the customers who purchase products and services. We can unify the nations around a light and hydrogen source of energy that will ideally be located in the most effective location. While this may sound like a distant future accomplishment, it is a reality we can largely achieve as a planet by 2050.

An ideal location for a Nuclear Fusion Reactor could be in Istanbul, Turkey, positioned away from populated areas while generating value and economic growth for the continent. While the location is uncertain at this time, I am confident that in the years ahead, we will find the solution by creating a reactor that hopefully relies more on light than transmission of electricity and other sources of energy, while strategically positioning it globally.

Whichever nation controls the majority of the supply of sustainable energy will control most global power. Currently, the largest producer of oil and gas is the United States of America. Evidently, the nation with the most power is the United States. The transition to hydrogen and light energy will determine who holds the most power in the future—hopefully, the near future. By practicing the Elemental Frequency Theory, we can optimize the distribution of elements across Earth, from the core to the land that supports our nations, all the way to the atmosphere that fills the air we breathe.

Oxygen can be created through vibrations, as well as other elements such as light. All elements can evolve. Evolving by creating larger atoms going forward, while also shrinking them down to foundational matter is important to the proper development of all space-time.

On Earth and within the solar system, it's important that we make the most of light. It is foundational to the Universe and what created the sun at its core, as well as the planet we call home. Controlling the future of energy starts with utilizing light and hydrogen globally, which will drive the market and the development of all products, as well as the degree of service that humans exemplify for others.

This will consolidate information, intelligence, interest, and impact into what I call the Four I's process. In a way, almost everything is information. All matter on Earth is information. Intelligence develops from information and data. Interest is what humans and even Artificial Intelligence have in content and matter to the point where their investment of time can maximize impact. Impact is the outcome that comes from taking interest in something, physical or spiritual. Regardless of whether it is made up of elements or just a spirit, the Four I's process can always be utilized to maximize the best impact.

The chemicals and frequencies that make up all matter drive the evolution of our planet. They also drive the evolution of every other planet that is habitable and every one that is currently viewed as inhabitable. As mentioned, the Universe began with light, and the amount of light we have universally is a fixed quantity.

It cannot be created or destroyed, so we must optimize the light we have. Regardless, I believe that light is continuing to expand our Universe from the Big Bang, or as many know it, the Beginning of Our Time. Light, or hydrogen as we know it, is made of one proton in the nucleus and one electron that orbits the atom. This composition is what the Universe began as, but it can evolve into larger, more proton-dense atoms as Matter, Area, Time, and Thoughts expand and develop.

We currently exist in what I believe to be a single Universe—the one we call home. A Universe can contract, expand, and even multiply into a new Universe. By doing so, we grow time, space, and matter. Naturally, this process is always sustainable because time, space, and matter trade between neighboring and new Universes rather than disappearing from existence completely.

When the Multiverse trades matter with a Universe, it increases the need for time and space. As time expands into the future alongside space, it creates further room for matter to exist and optimizes the frequency of existence. The frequency of matter is typically at its best when life properly exercises it through the mind, body, and especially the soul.

A key to properly exercising the Frequency Theory comes partially from religion and meditation. The best meditation I have practiced is Transcendental

Meditation, also known as "TM." The great thing about TM is that it requires no religious teachings or beliefs. Instead, TM focuses on guiding your thoughts toward your third eye, located in the pineal gland. The pineal gland can be accessed by focusing your thoughts between your eyebrows. With ease of breath and thought, you can reach inner peace while also engaging more easily with the matter of the Universe, optimizing the alignment between its elements.

This practice involves sensing the matching and pairing of elements within the Universe, particularly on Earth and within its atmosphere. Matching the vibrational frequencies of elements optimizes our resources while simultaneously expanding space, time, and matter within our Universe. Ultimately, growth is enabled by the individual's ability to match these frequencies, and the time-space-matter presently gifted to us serves as a foundation for creating a better and brighter future.

Don't just use your thoughts to matchmake elements and life forms—use your body, too. Your body, which also inhabits your soul, can feel what is best, not just think it. The body is a valuable tool in demonstrating frequency theory, just as the mind and soul are.

What's unique about frequency theory is that every person can practice it simultaneously. Eight billion humans can engage themselves and interact with every element in the Universe to extend time, space, and matter for all—at the exact same time. When many practice the theory together, it enables the ability to achieve optimization through the exploration of matchmaking elemental pairs.

The Universe has one total vibrational frequency, created from energy, time, space, and matter. Matter is what humans are made of. Within our lives, we experience time and space, which have the potential to last forever. Infinity is a possibility we can grasp.

Let's grip the future firmly as we create more time, more space, and more energy that matters to everybody. Aiming to birth a new star in our Solar System by utilizing a light and hydrogen network on Earth that uses the planet's gravity and technology to bring together gas and dust clouds of hydrogen and creates light from fusion of those atoms together. This could enable the creation of a second Sun in our Solar System. Enabling us with greater time and energy to thrive on our home planet, Earth. Energy comes in the form of elements, and these elements should be embraced to live our lives to the fullest. When I speak of living life to the fullest, I mean living as long as possible in the best way we can. That means making the most of the MaTts we have, which is matter, area, time, thoughts, and spirits.

Molecular Vibration

The Universe started from Love, which is energy, in addition to Matter, Area, Time, and Thoughts. Light and the energy it contained initiated an explosive reaction we call the Big Bang, which expanded and converted NoMaTt into more MaTts to facilitate universal creation. It also began with the vibrational frequency of an element that enabled life to become evident further into the future.

Translational and rotational motions are foundational to the vibrations of elements. This vibrational motion originates from the element's core, or what is known as its nucleus, and can be called translational motion. Translational motion occurs when the entire nucleus moves by the same exact amount, in the same direction, and within the same amount of time.

Universally, this motion can birth time-space-matter, depending on the molecular and vibrational motion of the elements. Rotational motion, on the other hand, occurs when an atom spins around its own axis, much like how Earth rotates on its axis based on the North and South Poles. The Earth's spin creates each day, enabling matter and life to prosper healthily. Earth also orbits around the Sun, advancing and maximizing the potential of all energy in its area.

Translational and rotational motions are fundamental to the behavior of elements and the subatomic particles that comprise them. By tapping into these vibrational motions, we can optimize the power of each element. When the motion of one element aligns with the motion of another and is optimally pairable, they should be brought closer together to manifest time-space-matter both globally and universally. This can create perfect phenomena and reactions among elements in the present, ultimately gifting us an infinite future.

Sometimes, people identify others and think or say, "I love their vibe." This could be due to the alignment of matter and spirit or simply the expression of molecular vibration within space-time-matter. Space represents where molecules and their vibrations reside. Time creates opportunities for space to expand and enables matter to form planets and stars. Matter allows all life to interact within space and time, making the most of their molecular makeup—not just physically and emotionally, but spiritually as well.

Similar vibrations are drawn to one another. People are often brought together through their similarities, especially their matters' commonalities. The elements that compose all our matter are what you and others feel in spirit. You can sense the molecular makeup of another when they are in relation to you—not just in direct proximity, but even at a distance.

Treat others with love and goodness, and the vibration of each molecule that comprises you will be enhanced, leading to a life that lasts longer and is perceived as better overall. Conversely, if you treat others poorly and unkindly, it won't just affect their molecular vibration, it will also affect yours too. The way you act determines both your present and your future. Take advantage of God's gift by making the most of the present and yielding the best in the future.

Embracing the Soul

A soul is the spirit of life. Collective consciousness is comprised of the collection of souls. Typically, the soul, along with love, is what drives friends and family together. Sometimes, when a man matches with a woman, it results in marriage. This brings together two individuals who can potentially birth the perfect child based on their soul alignment. In addition, the couple can embrace the ideal relationship between them.

Relationships typically come in the form of friendships. A collection of friends is what people call friend groups. It's best to be friends with those who are most like you because it enhances loving evolution to the fullest. When souls are matched, space, time, and matter are enhanced.

The most positive things in creation come from the mind, body, and soul. Aligning these three life components will enable us to live forever. It starts with the heart. Feel each beat. Drive the blood in your body to your mind. Connect mind and soul directly as you unify the two with your body.

Engage your thoughts in your "Third Eye," as

some call it. Your Third Eye is really your pineal gland, but it consciously connects everything through the spiritual realm. While spirituality isn't fully explored in science, it may soon be discovered through the connection of our minds. Your mind creates all your thoughts, and thoughts typically control the body. However, without a human source like the heart, there is nowhere for the idea to start transmitting blood. Love begins in the mind and is guided to every part of your body by the heart. Hormonal releases in your mind determine your body's function. When love is exemplified and reciprocated, it causes more hormones to be released and unifies you in that relationship. This will hormonally encourage you to want to be near them again.

Love and connect with all those around you, and your mind will evolve along with your cellular composition. If someone is hurtful to you and others based on how they think and act, love them to create change—or embrace them and all of humanity by giving them tons of love after all. This will not only change you but others as well. Consciously and collectively loving will evolve communities, both partially and wholly, in the end. Let's make that progress evident with the best outcomes.

Personally, I grew up in the most segregated community in the world, Elm Grove. In Elm Grove, there is hardly any diversity at all. Race is what drives communal separation in Milwaukee. While the communities in this beautiful city and many neighboring areas have been becoming more diverse, my childhood hometown has not. To maximize the value of life, diversity in race is best for the community, while matchmaking is still ideal.

As a planet, we can diversify now to create an optimal future. As neighboring cultures adopt a more unifying way of thinking, thoughts will be better guided toward a core truth. This truth can be recognized in education, economics, religion, and all aspects of life and the universe. Segregation and unequal division have brought me to tears. Differences can drive us apart, but color should not, just as our differences should unite us. When we focus on the one truth through collective consciousness, it will evolve and make us all better. Evolution will be maximized when we embrace love.

This was the most segregated suburb in the most segregated city, to the point where it brought me to tears. While differences can drive us apart, color should not. The way we think should bring us together so that each thought in the collective consciousness makes us better. Loving truth over falsehood will evolutionarily push you and others further.

The soul exists despite the lack of extensive scientific research into it. Although research on the subject is largely absent, the power of the soul is fully present. As mentioned previously, the soul and your spirit reside within your mind and body and exist within the collective consciousness of all life, matter, space, and time. Be good, and your goodness will stay with you for a lifetime. Your soul can be utilized in many ways, and you can even access specific memories stored within it. Make the most of time by understanding that your Spirit will last forever, no matter what. Even if Matter, Area, Time, and Thoughts collapse into a Black Hole, your Spirit will still remain.

A Perfect Universe

A Universe can last forever if all life, space, time, and matter always listen and embrace the one truth. For this to occur, the mind, body, and spirit of every insect, animal, and human must be in complete alignment with the collective consciousness. This possibility grows in likelihood when we believe it is not just possible but inevitable. The certainty of a perfect universe that lasts forever can be achieved when every element is in sync with what is best. "Universe" stands for "one song"—make sure you're listening to the right song, and your frequency won't be wrong if you're in sync with it.

Perfection is something almost everyone seeks deep down. However, discouragement can take away internal motivation, causing people to settle for less. Regardless, every human should strive to be their best. Achieve this by aligning the frequency of your mind, body, and soul. This will make you and others whole in the near term and into the future, hopefully for most of time. Molecular vibration within space can further this possibility, but it is nearly impossible to do without matter.

Breath is where much of the energy we receive fuels our body. Breathe deeply, breathe slowly, and breathe consciously. This will lead to a healthy heart, organs, mind, and a conscious soul that remains relaxed rather than stressed by circumstances. Stress may be constant, but it makes it difficult to reach perfection—whether as a person or a universe—if you are overwhelmed by it. Remember, make the most of all that is within you, and you will reach what is best for you and all else if you're committed to bettering them too.

Movement is another characteristic that should be utilized as human. It could be walking, running, or weightlifting—any form of exercise can be beneficial for the mind, body, and spirit. Some say they feel younger, but it almost always leads to better health. Physically, your health will definitely be enhanced. Mentally, your thinking will be clearer. Spiritually, you will feel better. With all of these benefits, it becomes difficult to say "no" to movement that is good for us. Trust in your mind and body as you move; you'll know when something is beneficial by paying close attention to the unity of your whole being.

Utilize your vision. We can see into the future with our eyes as we collectively and consciously create it alongside other life. It requires embracing molecular vibration to the fullest in order to achieve what is optimal for all of us. However, we don't need to exercise vision to create MaTt. What we need is the whole self, along with everything else. Along with those components, a matching of elements is required universally—especially on Planet Earth—and is completely necessary. This can result in an expansion of space and time and a shifting of matter.

Currently, we rely on oil, gas, coal, and electricity for much of our energy. While these resources supply us with sufficient energy, they do not attain sustainability. In fact, they lead us down a path that makes survival increasingly difficult due to climate change. Although the transition to a new energy source like hydrogen and pure light may be challenging, it has the potential to enable Earth to survive longer. I'd rather have a clean and sustainable planet than one that runs out of resources and ceases to exist within the next millennia. How can we achieve sustainability? Let's do it by making the most of light. Creating a secondary star. It's what our neighboring solar system is composed of, and it's what we can achieve by 2050 hopefully. If we get there, we can reach infinity—as Earth, our solar system, the galaxy, and the greater Universe.

Utilizing and Maximizing Light

Our light primarily comes from the Sun, but we still receive some light from neighboring stars. Light is such a fundamental energy source that it can power most of Earth's economy. Every day, light shines onto our planet, awakening us and supplying our crops with the energy they need to create food for all of humanity. Without light, Earth would not exist in its current form or be able to achieve what is optimal.

Light doesn't just determine Earth, our solar system, or a large part of our galaxy—it creates and enables the entire universe. It can expand and contract space, time, and matter over time. Making the most of light requires always relying on the light from stars to optimize each and every day. Night means taking a break from light in some regions of the planet. Most people sleep during the night to maximize the best utilization of each day, given that light is energy and darkness typically makes most people feel tired. This tiredness creates an ideal night of sleep because the energy in the environment directly affects the mind and body of each human.

Use the daylight cycle to your advantage. The frequency of the Sun dictates what occurs throughout our solar system. It guides humanity and all life on Earth while also enabling others to help and serve each other more thoroughly, rather than harming others or themselves over their lifetime. It doesn't just guide Earth—it guides Mars and other planets, too.

The Sun shines as a star that serves other planets in space as well. The race to create an optimal future can be tapped into by focusing on what is foundational. Light established our solar system, and then life occurred within that range, as Earth is the closest planet in the system to the Sun that can sustain it. We must make the most of what we have and create an even better and brighter future. Let's do more of what's right by making the most of light.

Serving Life Best

Each life form should be served in the best possible way. Cellular composition guides the present to create a better future, especially when it optimizes the use of resources for life on Earth and the greater Universe. Let's make the most of this opportunity by ensuring humanity serves as the best example of what the most evolved life form in our Solar System can accomplish. What we have created to date can serve as a foundation for the future. We must utilize what is readily available to manifest an optimal future. Infinite life is still a possibility, and we should aim for that by creating unlimited time through atomic alignment.

Matching human beings with other elements and cells in life that maximize their potential evidently extends time and enhances our existence. Mentally, this requires energy and focus to achieve. A collective effort, where individuals optimize their cellular vibration in both mind and body, will make this goal easier to attain. A person's cells are composed of molecules and atoms, and these smaller components can function at their best—along with all the rest—when they vibrate in the most optimal way.

Family members typically share the most similar vibrations. Friends often come second, as we tend to surround ourselves with those who resonate with us. Strangers, however, can also align with one another when they express appreciation for who the other person is and what they do. We are in a time when maximizing the potential of elements—especially here on Earth—must be a priority in order to preserve and expand the planet's prosperity. This applies not only to human life but also to animals and the environment, and it must be monitored to magnify global success.

To reach this level of achievement, various industries such as energy, technology, and medicine must be fully engaged. Let's proceed wisely, exercising caution in our investments while ensuring that we invest enough to yield positive returns. While these returns are often economical, they stem from a deep sense of optimism.

Typically, when you serve, you earn. This principle applies from companies to customers and extends to citizens of nations. The level of service you provide determines the amount you earn. I encourage earning enough—but not in excess—so that you can help your family, friends, and neighbors live their best lives. Alternatively, if you earn more, consider redistributing your wealth or increasing your donations. While everyone's net worth may differ, integrity remains an individual and collective virtue. By working as a team, we can generate both organic and inorganic integrity within life and all matter, aligning elements universally while starting on an individual and communal level. Focus on what is near before expanding your vision to what is afar—this pairing mechanism is optimal among both individuals and nations.

At an even greater level, the Frequency Factor influences the Universe through matter, area, time, and thought indefinitely. The goal is to propel the Universe, our planets, and stars toward infinity, ensuring that we never doubt our ability to reach it. However, some scientists have popularized the Big Freeze theory, which suggests that the Universe will eventually run out of energy to continue its expansion, resulting in a dark Universe.

It's not fully clear to us what occurs after matter is crunched into a Black Hole, but it will most likely create a Singularity and may lead to the formation of a new Universe. A Singularity collapses space-time-matter to a point so small that it is hardly recognizable based on its past existence. A Singularity—one condensed point of technological knowledge and data in technology—will also begin to be approached as Artificial Intelligence advances.

Ray Kurzweil wrote a book called *The Singularity Is Near*, emphasizing the buildup and condensation of technology to such an extent that it will drive markets, hopefully in a sustainable manner. Powering technology that enhances our lives is its purpose, but it is also leading to energy scarcity and environmental contamination. I presume we can solve the challenges of energy and technology, but doing so will require the full commitment of personnel from many companies and countries globally. If we succeed, appreciation won't just be expressed and reciprocated by us but also by extraterrestrial life.

We have brought humanity and alien life together in the past; let's form a relationship that we know will last. This can be achieved by optimizing elemental power, ideally by expressing their frequency in a way that enables all life universally. In the future, if done correctly, this could allow Earth to become a universal utopia. Utopia can be reached by utilizing resources in the best possible ways as we go about our daily lives.

As our Earth rotates, we enhance our time by using data and human ability to drive both the economy and the environment in harmony with space and matter—not just near our planetary home, but far beyond. From Earth to other solar systems in galaxies in distant areas of space, the matchmaking of elements will create time and enable the formation of matter. Service typically dictates what occurs. Words drive actions, which determine consequences. These consequences, whether small or large, shape our reality. Regardless, there are countless opportunities in the present that can gift us a better future—one that may leave us at our best if executed exactly right.

Producing Maximum Potential

Preservation and pairing of elements optimize productive potential and the birth of our and other universes with space-time-matter. Space, time, and matter that are not yet in existence but will be yielded by the positive pairing of elements can be called 'NoMaTt.' This occurs when matter and time don't exist yet in an area where there is no space but it is suitable for universal expansion.

Max Trautz developed a theory in 1916, while William Lewis followed suit unknowingly, later developing Collision Theory, which ultimately complemented Trautz's theory. My AI assistant, CoPilot, provided insight here: "Molecular collisions involve reactant chemicals colliding with one another. Effective collisions produce formations as an outcome, like stars and planets. Activation energy is the minimum amount of energy required for a reaction to occur. Lastly, reaction rate refers to the frequency of effective collisions." The more successful reactions that occur among elements in space and time, the more optimal our universe becomes.

Maximum potential in an elemental or molecular relationship can be assessed easily by life. Once the best alignment has been observed, thoughts, subatomic particles, molecules, and objects can be arranged so that they are situated near the most complementary elements or at least feeling the force of the most similar ones. This should occur to the point where it eliminates NoMaTt, creating space, time, and matter in the future, allowing the being or object to contribute to the universe's expansion.

Expansion as a distant addition to the Big Bang is preferable to experiencing a standstill into a Big Freeze, where all space, time, thought, and matter collapse to a single point. Let's choose our options intentionally to optimize outcomes rather than suffer consequences from negative ways of thinking, acting, or feeling. All of these factors can affect us, others, and the greater universe.

Make the Most of Space (Even Nonexistent Area)

Space isn't constant. It can be created from NoMaTt. To make the most of space, you must maximize the potential of your galaxy and solar system, including the greater Universe. Doing so means appreciating all matter by pairing the optimal elements together. This process expands and contracts the area surrounding the planet you occupy through thoughts.

Consider your way of thinking and acting when pairing elements on each planet, within their atmospheres, and across the greater solar system as an act of service. It helps us rather than harms us, and that is something all of us desire more of. Let's act as if we can create a real heaven rather than a negatively consequential hell. Doing so will benefit as many people as possible. Isn't that what humanity ultimately seeks—peaceful prosperity? That certainly seems to be the case.

Earth is a giant planet, but compared to others like Jupiter, we occupy only a minor amount of space. Earth's total surface area is over 190 million square miles. With the vast resources we have globally—from water to crops, energy, and transportation, along with much more—we can overcome any challenges that God directs our way. Infinity is a possibility, even if it means using space travel and rockets to reach a new solar system in a different galaxy. Stay optimistic, because your thoughts influence time, space, and even matter.

By counting on God, we can find a way forward. If Earth ever becomes uninhabitable, we can transport ourselves through space to colonize Mars, which appears to be the next most habitable planet in our solar system. This conclusion is based on the planet's elemental composition and available resources. Mars could guide present and future generations in establishing a meaningful life—a new home with water, food, energy, and a sun that shines just as it does on Earth. However, if the Sun is at risk of survival then we must begin transporting through space to another solar system that has habitable planets to ensure our survival as a species.

However, before leaving Earth, let's make the most of what we have here. We can do this by utilizing the resources near our poles more efficiently and building societies in Antarctica, Greenland, and other underutilized regions. Antarctica alone contains about 70% of the Earth's freshwater in ice. If we position a hydrogen nuclear fusion reactor in the most optimal area, while keeping it in sync with the Poles, we could generate and distribute hydrogen-powered light across the planet through a network called "LightNet."

We are in a crisis. The way we are using our resources will overheat the Earth and contaminate the environment or explode the Sun. Let's solve our greatest challenge—energy. Naturally, all matter is energy, and all elements matter, whether they are solid or composed of magnetic components.

We must take advantage of the North and South Magnetic Poles. This can be done by positioning a nuclear fusion reactor in synchronization with them and the rest of the Planet, transmitting its energy globally. The generation and transmission of energy are issues that you and I can solve together.

Let's begin with the completion of a nuclear fusion reactor. From Commonwealth Fusion to the

International Thermonuclear Experimental Reactor (ITER) and many other companies, there are promising leads in the strategic development of reactors, but we must build them in sync with our technology so that it is not just attainable, but also sustainable. However, we must ensure they are precise in building what is best. The engagement of a qualified intellectual property attorney is essential.

Nuclear fusion is at the center of LightNet. However, the development of sustainable desalination plants is also necessary. In addition, hydrogen plants—or what I like to call "HydroPlants"—will be instrumental in ensuring a sufficient and efficient energy supply while maintaining climate consciousness. Following the construction of the reactors and plants, transmission must be addressed to serve customers across the planet. We must bring companies together consciously, emphasizing cooperation over competition to ensure the Earth's long-term survival. Above-ground transmission seems best to me, but using a corporation like The Boring Company may be beneficial for underground solutions.

Our Sun has enabled humanity. It shines brightly for us every day and night as the Earth rotates around it. Utilizing the same energy—fusion—as the Sun presents an opportunity that must be approached with caution. We rely on this nearby star to sustain us, guiding light to our land and sea. The Sun grows crops and fuels all life, from you to me. Regardless, a fusion reactor must be completed as soon as possible, as long as it remains safe for all regarding its creation of electromagnetic radiation as an energy source and the greater network it effects on the planet. However, if scientific studies prove it is unsafe or unsustainable for the planet, we need to stop progress immediately.

In the meantime, we can begin constructing sustainable desalination stations and HydroPlants, as this is a crucial next step. These can be located along the coasts of California, Oregon, Washington, and elsewhere. Building these plants will be challenging but attainable. Let's focus on what brings our society to a point where we are the best we can be. This is a step we can embrace—not as a corporate race, but as a cooperative accomplishment that will benefit all on our planetary home.

Light already affects all life on Earth; let's fully utilize its power to make the present the best it can be. The rest of time can be near its best if we efficiently and optimally use light to do what is right—not only in our society but also across all elements, molecules, planets, and the beings they could support within our solar system and the greater Universe. However, light travels beyond our solar system, so let's not stop at the Earthly planetary perspective. Instead, let's extend our vision beyond Neptune, the last planet in our solar system.

There is a potential that the Sun might explode in the future leading to a blackhole, which is certainly a risk, but our thoughts can even influence its future development. Use your thoughts to align atoms Universally to the point where it enhances the life and power of the Sun substantially and sustainably. In order to communicate with the new life that will develop in the Universe using the **Alphabet** and **Numbers** with light.

Light is a force that can make things right, even when they start off wrong. Maximizing what is right is something I plan to do, and I'd love to do it with you, in as cooperative and beneficial a manner as possible.

Create Time

The brilliant scientist Albert Einstein made several universally impactful discoveries, one of which is that time is not absolute. As a matter of fact, time slows down for those traveling at faster speeds. This aspect of relativity extends to every planet and every being within our solar system, galaxy, and the greater universe. In the Special Theory of Relativity, Einstein also proved that time and space are intertwined in space-time, which is a four-dimensional continuum. In this continuum, matter can manifest to fill it with elements that may eventually lead to purposeful life, depending on elemental evolution. This is only part of the solution to optimizing the universe.

Time itself can be created through elemental matching, which leaves evolution at its best. Pairing the best together extends space and time and can even lead to the creation of matter by developing NoMaTt. This ultimately yields space-time-matter while creating a future initiated by the light extension of the neuro-network in the human mind and the matching and makeup of matter globally and universally from the heart and soul. It can be difficult to do, but it's necessary to lead to the best-case scenario with all that we know.

Human power can achieve what's best when it is cooperative rather than competitive, especially when that cooperation targets bettering life, health, knowledge, wealth, and more. Cooperation over competition leads to maximum production. The United States of America is the country I grew up in, and we have things pretty well off. Most Americans have food to eat, jobs to tend to, families, friends, and a whole plethora of other things that make our lives great. President Donald Trump said, "Make America Great Again," but does so in a way that maximizes competition with other countries, ultimately hurting us in the end. Cooperative advantages are better than competitive advantages because the cooperating parties can serve one another to an ideal point. Cooperation over competition leads to maximum production.

Elements almost always cooperate rather than compete at their atomic level, but some compete for space and time to try to forge a better future, although working together leads to things being best off over time. Elements matter, and matter is elements, while making the most of the present will gift the best future. Now is what we need, and now is what we'll always have, even as space, time, and matter react and bend through our universe. One universe, one song—or what I like to think of as frequency—which we all hear and see every day as we live our unique ways that suit us better cooperatively.

A World With or Without War

Without cooperation and with mostly competition among countries, companies, citizens, and even customers, conflicts can often escalate into war. If we cannot meet all people's needs, they may retaliate physically to secure the resources they require most. Some fight for power, while others fight for the security of their people. A world without war—one of peace and prosperity for all—is what I believe we can achieve.

War begins when information, in the form of matter, generates intelligence that is typically imbalanced, along with unequal resource distribution or direct conflict. One entity possesses more than another needs but is unwilling to share. This imbalance significantly affects their frequency of being and their relationships, but even more profoundly impacts their mind and body. I do not want to minimize its effect on the Spirit, however.

I witnessed this when I was just a child, a foreign country began planning 9/11 in collaboration with the Middle East. The foreign foe, at the time, instigated the attack on the Twin Towers due to inequality and an unproductive resource distribution both within and beyond its borders. However, this strategy did not serve them well. Today, they face major economic struggles and difficulty in serving their citizens due to ongoing competition rather than global cooperation—particularly with the United States. We are a supportive nation, and even more so when we are appreciated.

My mission is to save the planet, and I believe we can manifest a solution that will be generated within the next five years. This solution will ensure Earth's survival for as long as it can be. The network *LightNet* will take time to build, but by spreading light to every global economy—from neighboring continents to nearby cities and towns—*LightNet* can serve as its foundation. Light created our universe, powering our food production, and defines what we call day and night.

Let us optimize this first resource, hydrogen, to establish an atomic network across the planet—from the North Pole to the South Pole in Antarctica. We can then harness the Earth's magnetic field to foster a marketplace that efficiently locates, retrieves, and utilizes elements and molecules. By combining or keeping them separate in the most effective ways, we can control the ideal production of goods globally.

Magnetism is so powerful in elements that matter itself can even sense needs and desires within life—and life can reciprocate. Then, companies and countries can serve these needs, both for customers and citizens. Service is what builds the largest companies and the strongest nations. Let us serve wherever it is most optimal.

By precisely and equitably distributing goods and services on a global scale, we can likely prevent war entirely. We can create wealth, health, and good wellbeing for all life—universally.

Losing or Creating Time

Our biggest fear as the human race, overall, is probably that we'll end the Earth and ultimately lose time that we could have in the Universe as a planet. Although time and space are relative, many dedicate their lives to making the most of both. That effort by humanity has largely yielded beneficial results, but it has also led to significant problems, such as climate change and risking the one Sun we have and its ability to survive through technological advancement. Almost all people have contributed to climate change in some way—it's simply the way our society operates. To address this, our systems must be redesigned, from production to supply and, ultimately, to consumption.

If light does not continue to shine into the future and instead contracts toward where it began, we will lose time as a society to progress forward. Thankfully, almost every one of us has the ability to expand time by utilizing NoMaTt to create space, time, and matter for the future. This can occur through Element Matching, which perfects the Universe to a degree. Since there is no better than perfect, our focus should be on achieving perfection for Earth and the Universe.

This means unifying humanity to think together for the greater good.

All life has the ability to think to some degree. We can even adopt a technique that involves thinking with animals and base elements. This concept lies at the core of Elemental Frequency Theory—thinking with all matter to maximize its potential indefinitely into the future by pairing atoms in a way that extends space, time, and matter by utilizing the Creation Equation.

Maximizing Matter

Creating matter is necessary to continuously extend life into the future, as it encompasses NoMaTt. This can be achieved through the Elemental Frequency Theory's method, which involves sensing which atoms, molecules, animals, and human beings are most similar as you focus your thoughts on your pineal gland, also known as your third eye. The third eye can be engaged and optimized for you as long as you exist.

Scientists have identified the pineal gland near the front-center of the brain, while Mari Silva has located it between your eyes and right above your nose. That point is where you can direct your thoughts while allowing energy to flow to the pineal gland in your brain.

Anyone with a mind has the ability to pair atoms and molecules in a way that benefits all life on Earth and in our Solar System. Begin by practicing how to sense what elements naturally align around you. Then, implement the strategy with your third eye to bring together the most optimal matches. Keep in mind that NoMaTt will gradually disappear into the future as you and others pair particles. This process will foster the expansion of maTts even further into the distance, beyond our origin point.

I hope we continue to grow from that point forever, but achieving this will require more than just you and me. It will take a collective effort—a sum of beings—sensing through seeing, hearing, tasting, feeling, thinking, and creating to drive the Universe forward.

Discovery determines what can be created within. We continually discover more about space, but we need to make the most of what we have with the human race first. I have faith in our race and what we can accomplish—ultimately, living for eternity. At least, that's the big hope, but a hope that I know we can reach. Let's achieve it by making our lives better together through the implementation of the Elemental Frequency Theory.

The theory may sound complex, but it is quite simple to apply. Implementing it can save and preserve the human race. Ultimately, it introduces value to people—value they previously didn't recognize or appreciate—making them more whole than ever before.

Try it once: think between your eyes and sense the elements that are most attracted to one another. Pull that matter together through space, and you will leave an optimal universal mark in time—one that can be expanded upon.

How brilliant is that? As a human, you left your mark on the Universe by expanding space-time-matter using your thoughtful frequency. Continue to create more in place of NoMaTt, and I will continue to thank you for what you've done.

Matching MaTt—matter, area, time, and thoughts—creates an optimal Universe when it is synthesized in sync with all of its surroundings. Perfection among us can help lead to an infinite future if we continue to make the most of what God gives us.

Trust in God, and God will trust in us. That trust can endure for eternity if we exercise our intelligence in science, technology, engineering, mathematics, and more. Intelligence comprises all of MaTt, but it starts with our knowledge of matter, area, time, and ultimately, thoughts. Thoughts enable us to experience the best present moment. Use your thoughts wisely, and you can create an ideal future.

Making Area

An area formed and emerged in our Universe from a previous Universe that God occupied. God transported matter through a wormhole into our area so that we could live. Using maTt, God created within NoMaTt, bringing forth matter, area, time, and thought. This MaTt came from the past but was born through a hole that blossomed into a future opportunity for life and all elements.

For us, this has created a Universe with so much space and area that we can live forever and continue expanding. We may even be able to utilize Black Holes rather than fear them, as they contain a vast amount of condensed MaTt that can expand into NoMaTt, empty space, creating or expanding a Universe. In a Black Hole, MaTt is compressed into a single point. This point can embrace matter, you, energy, and spirits to more fully and perfectly manifest a Universe filled with peace, prosperity, love, and infinite opportunity.

Human emotions start with the mind and heart while displaying to all the love they feel. Love is ideal for evolution—or what I like to call a solution. Use

MaTt to exercise love for all matter in every area, at all times, with every thought. When love is met with love from one human to another being or element, it forms a key to the future by optimally shaping the Universe. This especially occurs when the best-suited elemental frequencies are matched with one another. Matching and pairing perfect elemental frequencies can perfect the Universe, even creating a new one at times.

Every Black Hole is comprised of MaTt, which can grow into a much larger Universe. Thoughts from the Matter can drive elemental pairs together, expanding the area within the Black Hole and developing more time to reform a part of a Universe. Many thought that a Black Hole is the end of their time, but utilizing MaTt can save the hopeless with thought and emotion in a Universal solution.

Evolution falls in line with this elemental pairing. Pair the best match with one another, and all the rest Universally can feel it too. This is what is called the Frequency Factor—the best frequency from an element can be made when it is in optimal positioning in relation to other MaTt. Einstein theorized that Energy is equal to Mass times the Speed of Light squared, and I believe this to be true when it encompasses area, time, and thought too. You and I are one from where we begin and where we could end for a brief time, but there never will be a true end for us. Instead, we'll convert and evolve moving through space with light and matter.

We must embrace truth and believe in it to make the most of MaTt. Doing this will lead us to feel, hear, and see all that we need—even into our deepest desires. Our needs and wants can be reached as one Earth, Solar System, Galaxy, and Universe. One verse, one song—let's do everything right, not wrong going forward, and we'll experience infinity on Earth.

More Time, More Opportunity

Time is a resource that can be created and utilized and we do this through the evolution of life. Elemental Frequency Theory optimizes time by matching matter within an area, allowing time to be used thoughtfully in the present and into the future.

This matchmaking begins with thoughts, which can produce actions that unify or separate particles on Earth, in space, and even on distant planets or stars like the Sun.

The creation of MaTts in our Universe happens simultaneously among all the resources, which is energy, a manifestation of spirits, matter, area, time, and thoughts. Whoever makes the most of MaTts will make the most Universally.

Your goal shouldn't be to maximize economically but to produce as much as possible in the most positive and optimistic way. Possibility leads to plausibility when the correct strategy and tactics are used to address and solve the situation facing that matter at that time.

Despite this, time is still relative, as Einstein

discovered, and this relativity leads to different experiences of time throughout area/space. Different experiences of time may lead to different optimal strategies depending on the matter and area.

This encourages supportive thoughts that strengthen the Universe as One. We don't want to divide MaTts from their optimal value in the current Universe or from what will be discovered when we convert NoMatt into MaTt—granting us Utopia through proper elemental pairing.

Infinite Thoughts

Thought begins with an input sensed through touch, taste, smell, hearing, or sight. These thoughts are then sent to the brain in humans or to a thinking center composed of elements and molecules in more simplistic life forms. Thoughts are processed by the molecular thinking center and the parts of that center that receive, process, and act on them.

Integration follows as the information is interpreted by the thinking center. In the brain, neural networks, along with memory and emotion, contribute to conscious awareness. In most other life forms, thoughts are stored within each cell to the extent that they can reproduce and multiply into the future, acting in a way that best ensures survival. This process is universally optimal for life.

Thoughts initiate actions because they shape what occurs on Earth and in the greater Universe. Practicing thinking like Yahweh helps bring about the best for yourself and everything around you. Yahweh signifies "I am" in a way that manifests what you declare. At times, you may need to repeat a thought multiple times until it resonates with you and aligns with everything surrounding you.

For example: "I am a Best-Selling Author with my new book *Creation Equation*." If the frequency doesn't feel true when I first think or say it, I should repeat it until the truth sets in within the larger Universe.

Living Forever

Infinite life is often a goal for humans, and it is especially a goal of mine. I want to help make the Universe last forever, which is a common aspiration for many. Naturally, once MaTts are formed, it does not die; instead, it can condense into a Black Hole as MaTt Points, but it will never disappear. The disappearance of MaTt does not occur—only a transformation.

Let's make the most of what we have by embracing MaTt with optimal actions within our Universe, beginning with the self. This will make forever better and achievable for us and all else. When searching for direction in life—whether in what to say, how to act, or how to feel—go to God. Ultimately, MaTts guidance will serve you well personally and universally.

Doing what MaTts need and want will lead to an abundant future. That is something all life is searching for. Since we all are related to God, let's do what God wants. It will also enable us to live infinitely. Achieving infinity is an endless mission because it requires us to work forever into the future. However, that does not mean we cannot attain it with sustainability.

MaTts comprises everything, so it makes sense to do what everybody and everything needs and wants to accomplish. That may mean acting like a society, as we do, to optimize time and thoughts (intelligence) so that we can reach an ultimate interest that serves us all, as well as the greater matter in an area of significance within our Solar System and Universe, leading to the achievement of greater potential. How we position ourselves and our thoughts leads to actions that comprise all that occurs throughout space—on planets, stars, and among life.

Some discoveries made by scientists have led some to believe that we may be able to live forever. A few of these include gene editing, cellular rejuvenation, nanotechnology, gene therapies, telomere extension, stem cells, and anti-aging drugs.

Gene editing utilizes specific techniques to modify genes, potentially correcting genetic defects and enabling a longer life for the patient. It's quite remarkable. We most certainly need more research in this area. This technology has the ability to alter a DNA sequence and influence a person's life forever.

Cellular rejuvenation is a process aimed at restoring unhealthy or damaged cells to a healthier state, effectively reversing the cellular aging process. Cellular aging is targeted in a variety of ways, including stem cell approaches, gene-editing technologies, epigenetic reprogramming, and mitochondrial restoration. Even physical exercise can slow stem cell decline.

Nanotechnology is something else scientists have thought may enable us to have infinite lives. Technology at the nanoscale deals with structures measuring 100 nanometers or less, particularly involving the manipulation of atoms or molecules, which is foundational to Elemental Frequency Theory.

Gene therapies are treatments that modify a person's genes to treat or prevent disease. The goal is to correct the genetic cause of a disease. How does it work? It replaces a faulty gene, inactivates it, introduces a new gene, or edits the genome. How is it performed? It can be done **in vivo** or **ex vivo**—either through an injection directly into the body (**in vivo**) or by removing cells from the body, treating them with genetic material, and then transplanting them back (**ex vivo**).

Telomere extension is the process of rebuilding the protective caps at the ends of chromosomes. These caps, called telomeres, are made of DNA and proteins. As telomeres shorten, lifespan generally decreases, but telomere extension may offer a solution to counteract this process.

Stem cells are being studied in meaningful ways to extend human life. The body contains many different types of cells, but there are two main types of stem cells: embryonic stem cells and adult stem cells. Stem cells have the potential to develop into bones, heart muscles, nerves, and other organs and tissues. Using cellular structures and magnetism in a light-based machine could be the key to unlocking eternity for human life. This approach could transform the body into a healthier state, making the idea of living forever seem within reach.

Anti-aging drugs are precisely what they sound like—medications that slow down the aging process. These drugs work by repairing and biologically enhancing the brain and body in alignment with an individual's soul.

By utilizing these strategies and tactics, we can extend life far into the future. Doing so within an infinite universe makes the concept of eternity no longer seem impossible. Instead, it becomes an achievable possibility—provided we focus on what we truly need and desire, ensuring that our wants remain healthy.

Matching or Breaking the Bond

Elements have relationships, and not all that is matter is equal. This applies to the beginning of our Universe, where Hydrogen atomically evolved and brought in matter from a neighboring Universe, such as Helium and even Oxygen. Eventually, this light condensed by being matched with other atoms—so much so that stars were created early on in our planetary existence. Then, 9.2 billion years after the beginning, the Sun came to be.

Every day, you look up at the sky and take in the energy from the Sun. Be thankful for what God has given us with every breath and word you make. It's incredible to be able to make sense of the past in a way that enables humanity to live infinitely into the future. It is within reach, and I'm hoping *Elemental Frequency Theory* will teach you a way to live your life that benefits not just you but everybody.

Collectively, through cooperation, we can achieve our ultimate destiny—being our best selves as a collection of countries and citizens that represent truth in whatever political practice they administer, whether it's Capitalist or Communist. Focusing on optimizing each citizen's and candidate's talents globally will most certainly yield success. Rather than launching a war between the United States and another country, working together is undoubtedly the better path to serving each citizen within both nations.

Humans dictate most of what happens within the *Elemental Frequency Network* on Earth. I'm not just speaking of the Internet, cable, or radio networks. I'm referring to how our mind, body, heart, and spirit dictate thoughts, feelings, and actions, which administer a frequency from our very beings that vibrate throughout society—beginning from the cells in our body. Align your mind, body, heart, and soul with the truth always, and recognize that a fault is not the end. Rather, it can be a point of practice on your way to perfection—where you live optimally.

Optimization is achieved through exploration, with direct attention being brought to what is best. Ignore all the rest. You have 24 hours in a day—match your time with the matter you are (body), in an area (where you are), with your thoughts (mind). If things become misaligned, then break the bond and find a better match between elements, cells, and molecules. Start with a thought that will power you forward, beginning with your mind. Evolutionarily your best match will remain the same infinitely with love and evolve into the future.

Going from Good to Great

We are all conceived by sperm and egg, receiving life from both a mother and a father. We begin as a point of perfection, with the potential to evolve optimally from there or make the first choices of life—whether good, great, or optimal decisions. These decisions can serve our early friends, our family, and eventually, everyone in society. Many times, our choices are shaped by our parents, especially while we are still in the womb.

Living a perfect life would create a sense of infinity in our existence, which happens to be one of my life goals. However, true infinity must be reached collectively, forming a web of belief that humanity is destined to achieve. By making the most of Earth and the entire Solar System, we can hopefully enable the Sun to last as long as possible as well. Elemental Frequency Theory suggests that a long-lived Solar System and Sun are possible, while creating a new star within the system is a possibility too.

Achieving this requires matching elements with their most optimal counterparts in the Solar System using magnetism, frequencies, and energy. These forces drive atoms together, creating molecules, separating them, and even altering the atomic composition of atoms. When this process is carried out wholistically throughout the Universe—with a concentration on our Solar System and the galaxy we belong to—it draws Matter, Area, Time, and Thoughts from our nearest neighboring Universe. This expands our own Universe, creating MaTts in the areas of NoMaTt until all elements have a perfect frequency match. It is difficult, but it is possible.

What I wonder is this: Do past mistakes define our future, or can we still enable an infinite future through elements, molecules, and cells? I believe an infinite future is still possible, but I am not entirely certain. One way I plan to prove I'm right is by creating it.

Every element starts off in a good state and can evolve toward greatness—or deteriorate. Elemental evolution depends on the frequency that atoms emit among themselves and those around them. This process creates new atomic makeups or molecules that interact optimally with others. The relationships between elements drive our Solar System every day, especially here on Earth.

The number of living elements on Earth is abundant, but they must be fully appreciated and supported to sustain global elemental balance.

Supporting and aiding these elements throughout their existence—by matching them optimally with other atoms within our Solar System—will further enable our Universe to reach a state of perfection and, hopefully, an infinite future for every component.

This is something we have the power to determine. Let's use science to understand it, then push beyond what we once thought was impossible—until we believe that perfect elemental unity can truly be achieved.

Perfection in a Reflection

A new Universe can be formed when an existing one looks into the prior and matches its atomic makeup perfectly in sync and can achieve being better off by more efficiently engaging with atomic alignment. This perfection creates a reflection that adds another Universe into the future by making the most of the present God gave us in our current and prior Universes. It doesn't have to be a perfect reflection of the previous Universe, but it does have to match every element in the current one exactly right.

Space is abundant, but we must fill the Area with Matter that creates Time through its Thoughts and their frequencies to make the most of it.

This outlines and defines our beginnings as a Universe and the next one we aim to create in the future. Synchronizing all elements based on their frequencies within a Universal window and matching them molecularly is key to making the steps of creation.

Creation can continue infinitely into the future from where we currently are. To do so, we need maTts as well as NoMaTt. The Matter, Area, Time, Thoughts, and Spirits will expand into the NoMaTt, growing the Universe and perhaps creating another one if the reflection from the matchmaking is perfect.

Optimal elemental frequency alignment Universally is foundational, led by the one who embodies mind, body, and spirit. This leader serves the whole Universe.

Building Better Together

We can build badly, better, or ideally. This construction of the economy began billions of years ago when life first emerged. The evolution of life can even develop ideally, occurring within single-cell organisms up to mating beings.

In order to optimize the building as we do as a society, we must make the most of frequencies—bringing a sensation to the molecular vibration of life and resources to the point where molecules maximize the positive production they can generate from that connection, along with the remnants of all others. We must also optimize our use of light from the Sun, our nearest star.

Much of humanity's technologies offer some economic enhancement, but ultimately cause damage that harms us more than it helps. This is especially relevant to our networks, such as cell phone networks, the Internet, radio networks, and even cable television. These frequencies are harmful to elemental growth across the planet and even extend beyond, affecting our solar system and the one Sun we have. Many of our networks cause negative gene mutation and may lead to cancer over time.

We must transition to making the most of light and hydrogen to construct a Light Network. LightNet can power society with energy from cables and fuel produced by power plants.

We have an abundance of hydrogen in our oceans, and if we desalinate the water while molecularly breaking it apart, we can use the hydrogen for energy and light production, oxygen for the air, sodium for energy, and chlorine to maintain cleanliness of the water.

Saltwater is one of the most valuable molecules on the planet due to the essential elements it contains. The challenge lies in desalinating it with efficient energy use and then breaking apart the water using pure sodium and freshwater to extract pure hydrogen and oxygen.

Over time, we can utilize hydrogen for energy and oxygen for the air. As these elements are used throughout society, the Earth's magnetic poles can influence global weather, controlling the formation of water vapor and determining where it will fall as rain, ultimately returning to the ocean.

The sodium-ion batteries created from pure sodium can store the energy generated during molecular breakdown. This process will gradually convert more oceanic water into freshwater, reducing the proportion of saltwater. Optimism about our planet should be embraced concerning every element to ensure the most effective evolution.

Peace through Pieces

Elements make up nearly everything everywhere; they all matter because they comprise all matter universally. Often, elements work together to evolve everything around them, creating better circumstances.

Creation yields elements that reflect perfection from another dimension into ours, compounding to construct and evolve a new future in a new universe by matching elemental pieces peacefully together in the best way possible. Ideal possibility is something that all matter seeks. This could enable a utopia on Earth, which we could use as a foundation to populate our solar system, starting with neighboring planets that have magnetic poles and water, like Jupiter.

First, we must achieve peace among nations and all people on the planet. This typically starts with addressing humanity's needs and wants.

While these may differ between societies, they largely revolve around resources and their distribution among cities, states, and countries. Global distribution is closely monitored, and some very valuable resources are:

- Oil
- Steel
- Coal
- Hydrogen

Many of these resources vary in pricing based on demand, which shifts from one community to the next depending on the makeup of the people. People typically strive to support their well-being or create value in society.

Trading these goods is one way to do so and has led to the creation of many fortunes worldwide. I believe that many of these fortunes should be redistributed to those who need them most and are positioned to contribute value of their own. Each person has a unique talent that allows them to add something meaningful in their own way. Focus on what leaves you and others at their best.

Currently, we need to achieve peace between Russia and Ukraine. This requires the continuation and eventual conclusion of negotiations between President Putin and President Zelensky. Peace is within reach.

We don't have to watch soldiers die to redistribute resources to those in need. Instead, we must act to ensure the well-being of every citizen as part of our global responsibility. Service begins with candidates who provide governance for citizens, companies, and customers.

It's what I call the 5C Model. Countries have candidates who serve citizens and companies that deliver goods and services to customers. Customers' purchases of goods and services are taxed, and that revenue byproduct flows to the government. Then, candidates receive payment from tax income, while companies generate income from the profit they receive after customers purchase a good or service. Companies then pay taxes based on the income they generate.

War has broken out many times throughout history regarding resource distribution or international harm, whether it was the World Wars, the War on Terror in the Middle East, or any number of other conflicts.

Peace comes from individual people, who are made up of pieces, and those pieces can be optimally aligned to achieve peace spread across each nation that makes up Earth. Always do what's right, and the perfect reflection will generate a new universe.

Studying wars is like studying conflict across the planet, and those conflicts typically leave people worse off because there is a misalignment of matter—sometimes such a significant misalignment that it ends in disaster, like the atomic bombs dropped during World War II.

The Civil War was another memorable conflict that ultimately helped resolve much of the slavery issue and led to a redistribution of resources within the United States of America.

The deaths in war are always tragic; let's protect against them by not entering a war in the first place. That means not viewing humanity as a competitive race but as a cooperative game that can result in everyone and every nation being a winner. This is especially true when each person speaks and acts as if they are their best self, which is certainly possible. Let's make the best possibilities probable by how we behave infinitely into the future while matching in pairs with all those who benefit from a relationship between us and them.

Forming Genes

Genes come together to form the building blocks of life. Each gene, which makes up cells, represents the smallest unit of life that can exist on its own. These genes combine to create a single life form. Genes are made of DNA, which essentially serves as an instruction manual containing the information needed to create an organism. DNA is organized into structures called chromosomes, which are located in the nucleus (center) of every cell.

Humans have 23 pairs of chromosomes, totaling 46. They inherit one set of chromosomes from each parent.

Different beings pass on their genes in various ways. Evolution enables organisms to adapt and improve over time, shaping them into better versions of themselves. Much of this process is influenced by love. Genes determine traits such as intelligence, eye color, skin color, height, nervous system function, and much more. They act as the workhorse of the cell, driving genetic development as the organism evolves to ideally become the best version of itself.

For humans, this development occurs in the womb. For other species, it also takes place in the womb. Evolution continues after birth, but genes themselves do not change—those are largely fixed at birth. However, optimizing cellular function is possible through nutrition, lifestyle choices, emerging therapies, supplements, and environmental factors. These elements are fundamental in helping an organism achieve its best potential.

Cells are not exclusive to humans; they exist in bacteria, insects, larger mammals, plants, fungi, and all life forms on Earth.

We must work together to create optimal living conditions through the environments we build and the resources we use. This approach can lay a strong foundation for both the present and future of our planet.

Atomic Alignment

Each atom is best suited to a specific position or in relation to another atom or molecule. We should use our thoughts and their frequencies, along with our bodies and minds, to align each atom ideally with a pair.

This will drive both universal development and atomic development within those atoms that are molecularly paired with one another. Every atom has a frequency and can be identified individually. Each element is best matched with one or a few other elements within the universe. This matching creates the best circumstances within a given area and even influences the future. We can shape the near-term future by applying atomic alignment on Earth and within our solar system.

Practice using your optimal frequency in harmony with your thoughts, mind, body, and spirit. This frequency can position elements ideally. By working from this ideal state, we can create a future of perfection and utopia on Earth and beyond.

www.ingramcontent.com/pod-product-compliance
Lightning Source LLC
LaVergne TN
LVHW050921210926
848482LV00012B/492